Platypuses

Elsie Olson

Big Buddy Books
An Imprint of Abdo Publishing
abdobooks.com

abdobooks.com

Published by Abdo Publishing, a division of ABDO, PO Box 398166, Minneapolis, Minnesota 55439. Copyright © 2025 by Abdo Consulting Group, Inc. International copyrights reserved in all countries. No part of this book may be reproduced in any form without written permission from the publisher. Big Buddy Books™ is a trademark and logo of Abdo Publishing.

Printed in the United States of America, North Mankato, Minnesota
102024
012025

THIS BOOK CONTAINS RECYCLED MATERIALS

Design: Elena Klinkner, Mighty Media, Inc.
Production: Mighty Media, Inc.
Editor: Liz Salzmann
Cover Photograph: Clive/Adobe Stock
Interior Photographs: Ash/Adobe Stock, pp. 5, 7; Clive/Adobe Stock, pp. 17, 19 (top); Dave Watts /Alamy Photo, p. 21; ice_blue/Shutterstock Images, p. 27; Joaquin Corbalan/Adobe Stock, p. 19 (bottom); Jurgen Freund/Nature Picture Library/Alamy Photo, p. 23; Kryvosheia Yurii/Shutterstock Images, p. 25; Lukas/Adobe Stock, pp. 9, 18; 169169/Adobe Stock, p. 11; phototrip.cz/Adobe Stock, p. 13; slowmotiongli/Adobe Stock, p. 19 (middle); stas111/Adobe Stock, p. 14 (compass rose); tanarch/Adobe Stock, pp. 14–15 (maps); Willpembrokephoto/Adobe Stock, p. 29
Design Elements: flovie/Shutterstock Images (patchwork pattern); Mighty Media, Inc. (series logos & icons)

Library of Congress Control Number: 2024938325

Publisher's Cataloging-in-Publication Data
Names: Olson, Elsie, author.
Title: Platypuses / by Elsie Olson
Description: Minneapolis, Minnesota : ABDO Publishing, 2025 | Series: Odd but adorable animals | Includes online resources and index.
Identifiers: ISBN 9781098295141 (lib. bdg.) | ISBN 9798384915195 (ebook)
Subjects: LCSH: Platypus--Juvenile literature. | Egg-laying mammals--Juvenile literature. | Animals--Australia--Juvenile literature. | Monotremes--Juvenile literature. | Freshwater animals--Juvenile literature. | Curiosities and wonders--Juvenile literature.
Classification: DDC 599.29--dc23

Contents

A Platypus Encounter 4
A Closer Look 6
Water Wonders 8
Electric Animals 10
Down Under Dwellers 12
Location Station 14
Happy Hunters 16
Creature Feature 18
Eggs-cellent Mothers 20
Growing Up 22
Critter Culture 24
Threats and Hope 26
Odd or Adorable? 28
Glossary 30
Online Resources 31
Index 32

A Platypus Encounter

You're out for an early morning walk along a slow-moving Australian creek. Suddenly, you spot a sleek, dark shape gliding through the murky water. The creature pops its head out of the water. It has a duck-like bill and feet, a beaver-like tail, and otter-like fur. You've just met the odd but adorable platypus!

Platypuses are also called water moles, duckbills, and duckmoles.

A Closer Look

Platypuses are a type of **mammal**. They have thick, waterproof fur. It is usually dark brown to reddish brown. Adult platypuses are 14 to 25 inches (36 to 64 cm) long. They weigh between 1 and 6 pounds (0.5 and 2.7 kg). Male platypuses are bigger than females.

Platypuses can live longer than 20 years.

Water Wonders

Platypuses are **amphibious**. This means they spend time both on land and in the water. Platypuses have long, lean bodies. This helps them glide quickly through the water. Their webbed front feet act like paddles. A broad, flat tail helps them steer. Platypuses even have special folds of skin. These cover their eyes and ears to keep water out.

A platypus can increase its body heat to stay warm in cold water.

Electric Animals

Platypuses have soft and rubbery bills. But a platypus bill has a secret superpower. It is filled with thousands of special cells. These cells send out electrical signals. Platypuses use these signals to move through dark waters and find food.

A platypus uses its bill to catch worms and other prey. Its bill is so sensitive that the animal can hunt with its eyes, ears, and nose closed.

Down Under Dwellers

Platypuses are found only in eastern Australia and Tasmania. They live near freshwater creeks, rivers, and ponds. Most platypuses live in **rainforests**. The climate there is usually warm and wet. But some platypuses live in mountainous areas in Tasmania. It is cooler there. Platypuses' thick fur helps keep them warm.

Some platypuses live in wetlands, such as marshes and swamps.

Location Station

INDIAN OCEAN

AUSTRALIA

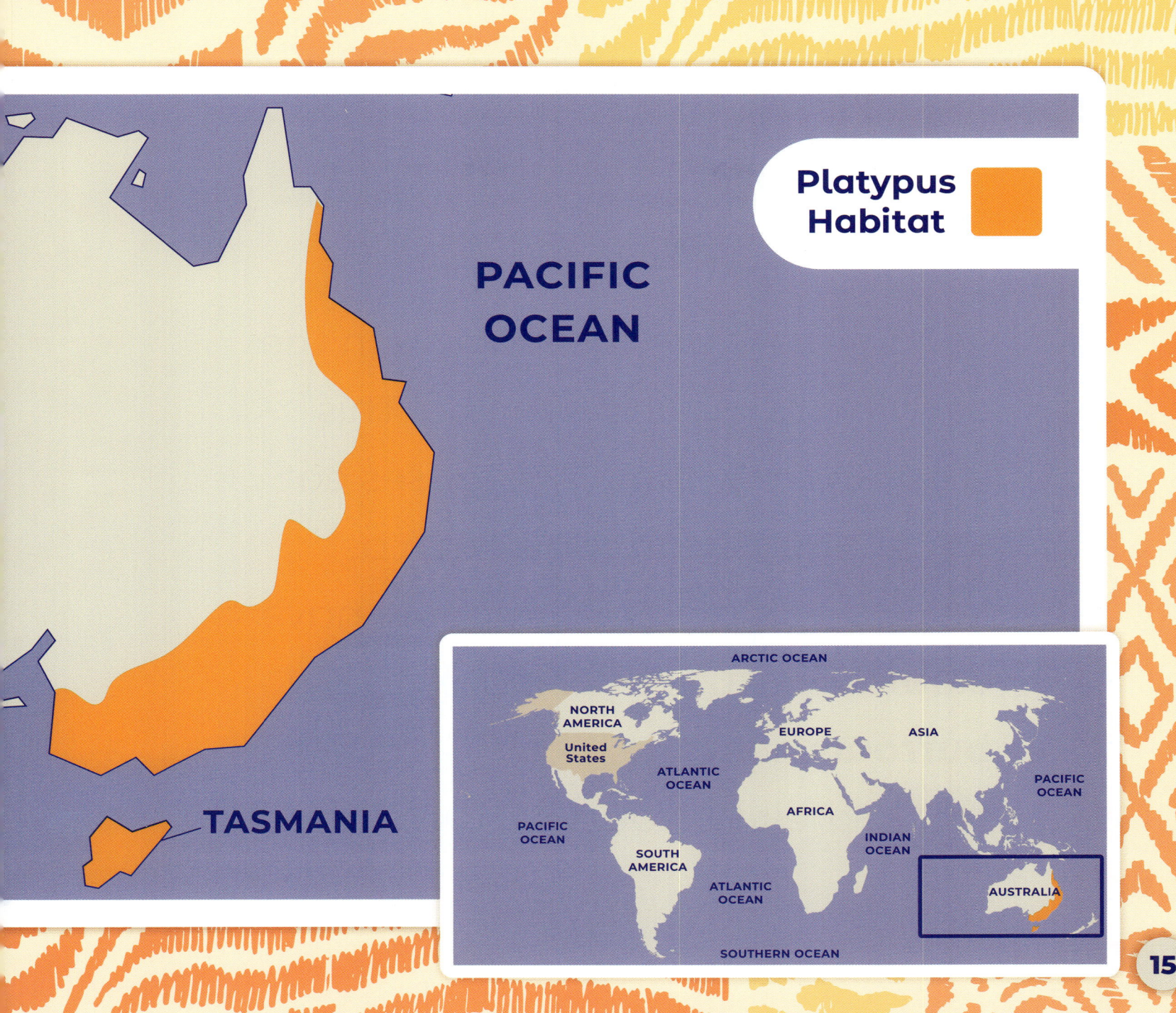

Happy Hunters

Platypuses are most active in the early morning and evening. This is when they hunt. Platypuses eat small **aquatic** animals such as insects and crabs. Platypuses have claws. They use their claws to dig **burrows** near the water. Platypuses rest in their burrows between meals.

A platypus's claws are long and strong.

Creature Feature

Platypus means "flat-footed" in Greek.

Aboriginal Australians have many names for the platypus. These include boondaburra, mallingong, and tambreet.

A platypus's foot webbing **retracts** when it is on land, revealing claws.

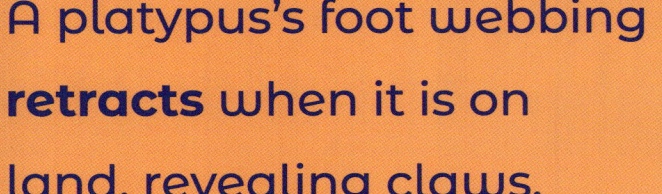

Platypuses don't have teeth. When eating, they use their bills to scoop up small bits of rock along with small animals. The rock helps them "chew."

Male platypuses are one of the only **venomous mammals**. **Spurs** on their back legs can sting predators and deliver the venom.

Eggs-cellent Mothers

Platypuses spend most of their lives alone. But males and females meet up once a year to mate. Platypuses are monotremes, or **mammals** that lay eggs. A platypus mother digs a deep **burrow**. She lays one or two eggs in the burrow. The eggs hatch after about ten days.

A platypus leaving its burrow

Growing Up

Baby platypuses are tiny and helpless. They spend their first months in the **burrow**. Platypus mothers produce milk for their babies to drink. After about four months, platypuses can swim on their own. When they are strong enough to hunt, they are ready to leave the burrow.

Some people call baby platypuses puggles.

Critter Culture

Platypuses are important to Australian **culture**. Many children's books and TV shows feature platypuses. **Aboriginal** Australians have many stories about them. In one story, different animals invite the platypus to join their special groups. But the platypus decides not to join any group. It chooses to stay friends with all the animals.

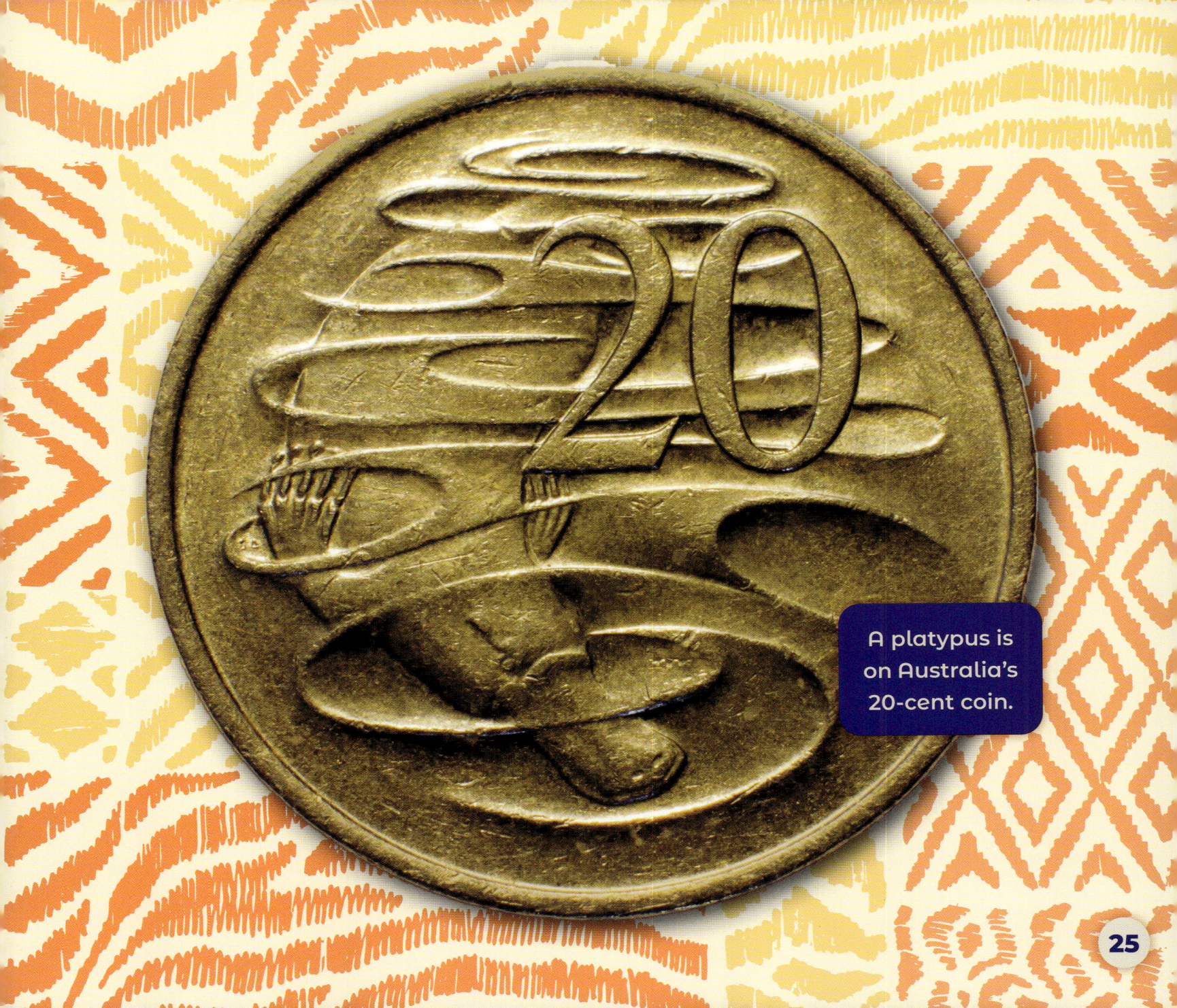

A platypus is on Australia's 20-cent coin.

Threats and Hope

Owls, eagles, and snakes hunt platypuses. But humans are the greatest **threat**. Platypuses get tangled in litter such as fishing lines. **Habitat** loss is another danger. Humans are building more homes and businesses where platypuses live. But there are people working to protect platypuses. They help platypuses stay safe and healthy.

Wild dogs also hunt platypuses.

Odd or Adorable?

Platypuses are unusual and beloved animals. They look strange, but they're also considered cute by many people around the world. What do you think makes an animal odd? What makes an animal adorable? Do you think platypuses are odd, adorable, or both? Why?

Platypuses are usually shy. This makes it difficult for scientists to study them!

Glossary

Aboriginal (a-buh-RIHJ-nuhl)—of or relating to the first or earliest-known people to live in Australia.

amphibious (am-FIH-bee-uhs)—living both in water and on land.

aquatic—living in water.

burrow—an animal's underground home.

culture (KUHL-chuhr)—the arts, beliefs, and ways of life of a group of people.

habitat—a place where a living thing is naturally found.

mammal—an animal that makes milk to feed its babies and usually has hair or fur on its skin.

rainforest—a tropical woodland with a lot of rain.

retract—to pull back.

spur—a sharp, pointed part.

threat—something that could be harmful.

venom—a poison made by some animals and insects. It usually enters a victim through a bite or a sting. Something that makes venom is venomous.

Online Resources

To learn more about platypuses, please visit **abdobooklinks.com** or scan this QR code. These links are routinely monitored and updated to provide the most current information available.

Index

Aboriginal Australians, 18, 24
Australia, 4, 12, 14, 15, 18, 24, 25

babies, 22, 23
bills, 4, 10, 11, 19
books, 24
burrows, 16, 20, 21, 22

cells, 10
chewing, 19
claws, 16, 17, 19
culture, 24

eggs, 20
electrical signals, 10

feet, 4, 8, 17, 18, 19
food, 10, 11, 16, 19
fur, 4, 6, 12

habitats, 12, 13, 14, 15, 16, 20, 21, 22, 26
hunting, 11, 16, 22, 26

legs, 19
length, 6, 8

mammals, 6, 19, 20
map, 14, 15
mating, 20
mothers, 20, 22

names, 5, 18, 23

platypus (word), 18
predators, 19, 26, 27
skin, 8
stories, 24
swimming, 4, 8, 9, 10, 11, 22

tails, 4, 8
Tasmania, 12, 15
threats, 26, 27
TV shows, 24

venom, 19

water, 4, 5, 6, 7, 8, 9, 10, 11, 12, 13, 16
weight, 6